Nature's Habitats
IN THE DESERT

Annabel Griffin

Illustrated by Rose Maclachlan

First published in Great Britain in 2024
by Hungry Tomato Ltd
F15, Old Bakery Studios,
Blewetts Wharf, Malpas Road
Truro, Cornwall, TR1 1QH, UK

Copyright© 2024 Hungry Tomato Ltd

No part of this publication may be reproduced, stored in a retrieval system, or transmitted in any form or by any means, electronic, mechanical, photocopying, recording, or otherwise, without prior written permission of the copyright owner.

A CIP catalogue record for this book is available from the British Library.

ISBN 9781835693544

Printed in China

Discover more at:
www.hungrytomato.com

Psst! I'm hiding on every page. Can you spot me?

CONTENTS

In the Desert	4
Majestic Beasts	6
Keeping it Cool	8
Feathered Friends	10
Small but Mighty	12
Brilliant Burrowers	14
Growing Gets Tough	16
Mighty Mammals	18
Where in the World?	20
Did You Know?	22
Who was Hiding?	23
Glossary	24

Words in bold capital letters **LIKE THIS** can be found in the glossary.

IN THE DESERT

There is so much to see in these dry and scorching places. Can you spot the animals and plants that make their home in the desert?

MAJESTIC BEASTS

These large animals have all ADAPTED to survive a long time without water.

Got the hump?

Camels can survive for up to 15 days without drinking. They store fat in their humps which helps them go longer without water.

Bactrian camels have two humps.

Bactrian Camel

Dromedaries only have one hump.

Dromedary/Arabian Camel

Baby camels are called calves.

A rare sight
The addax is ENDANGERED. There are only a few left living in the wild.

Addax

Desert Elephant

Long distance travellers
In the desert, elephants will walk up to 50 miles a day in search of food.

KEEPING IT COOL

REPTILES are COLD-BLOODED, which means they can't control their body temperature. Most of them spend a lot of time in BURROWS to try to keep cool.

A super long lie-in!
Box turtles HIBERNATE for over half the year.

Desert Box Turtle

Shaking off danger
Rattlesnakes use their rattles to make loud noises to scare off PREDATORS.

Rattlesnake

Mighty monsters
These are the largest lizards living in the United States.

Thirsty?
Desert tortoises can survive a whole year without drinking any water!

Desert devil
This strange looking creature is covered in spikes to help protect itself.

FEATHERED FRIENDS

Take a look at these amazing desert birds! They all live and act in very different ways.

Clever nesting
Gila woodpeckers often make their homes in cacti.

Gila Woodpecker

World's largest bird
Ostriches can grow up to 2.7 metres tall.

Ostrich

Tasty leftovers
Vultures are **SCAVENGERS**, which means they eat animals that are already dead.

Lappet-faced Vulture

Night owls
Great horned owls are **NOCTURNAL**. They sleep in the day and hunt at night.

Great Horned Owl

Kicking up dust
Roadrunners are super speedy and can run up to 27 miles per hour.

Roadrunner

SMALL BUT MIGHTY

These desert creepy-crawlies all have unique skills to help them survive in difficult conditions.

Big biters
Camel spiders use their scary jaws to attack bigger PREY, including lizards, snakes and birds!

Camel Spider

Dung Beetle

Poo pushers
Dung beetles feed on the poo of other animals. Yuck! Some of them roll dung into balls, like this one.

Greedy guzzlers
Locusts can form enormous swarms and travel huge distances, eating every plant in their path.

A sting in the tail
Scorpions use the VENOMOUS sting at the end of their tails to hunt their prey.

Quick silver
These ants are covered in shiny hairs that reflect the sun, helping to keep them cool. They are also the fastest ants in the world!

BRILLIANT BURROWERS

It's cool to hang out underground! Many animals escape the heat of the desert by digging burrows to live in.

This meerkat is on guard duty, looking out for predators.

Meerkats

Pack members
Meerkats live together in groups called *packs*. They all share jobs between them.

Burrow for one
Bilbies like their space. They usually live alone and will have up to 12 burrows each!

Bilby

Sand squatters
These owls like to take over burrows made by other animals.

Burrowing Owl

The better to hear with
These big ears are great for listening out for prey.

Fennec Fox

Feisty felines
Sand cats may look like cute pets, but they are tough enough to handle extreme desert conditions.

Sand Cat

15

GROWING GETS TOUGH

Most plants need lots of water to grow, but these plants don't mind dry weather.

Cacti

Clever cacti
Cacti come in lots of different shapes and sizes. They store water in their stems and are covered in sharp spines.

Pretty but prickly
This funny looking cactus produces pretty flowers and fruit that animals can eat.

Prickly Pear Cactus

MIGHTY MAMMALS

These MAMMALS are all tough enough to stand the heat of the desert.

Mountain Lion (Cougar)

Look behind you!
Mountain lions like to sneak up on their prey, often jumping down on them from above.

Butting heads
Male bighorns will use their large, curly horns to fight each other, to prove who is strongest.

Bighorn Sheep

Rare runners
These rare gazelles are very good runners. They can reach speeds of 50 miles per hour.

Arabian Gazelle

Smells like dinner!
Coyotes have a great sense of smell to help them sniff out food from a long way away.

Coyote

Red Kangaroo

Big bouncers
Kangaroos don't walk, they jump! They have strong back legs to help them bounce over long distances.

WHERE IN THE WORLD?

NORTH AMERICA

SOUTH AMERICA

KEY

= Desert Regions

This map shows all of the desert areas in the world. Not all deserts are the same. Different plants and animals are found in each one.

DID YOU KNOW?

Birds on the ground
Ostriches can't fly like other birds, but they can run really fast (up to 43 miles per hour)!

Mistaken identity
Camel spiders are not actually spiders, although they are related to them.

Strength over size
If you compare size with strength, dung beetles are the world's strongest animal. They can pull up to 1,141 times their own bodyweight!

WHO WAS HIDING?

Did you spot the little Egyptian tortoise playing hide-and-seek in each desert scene?

These tortoises live in desert areas in Egypt and Libya in North Africa.

They are very small and only grow to between 8-12 cm in length.

I'm very rare!

You would have a hard time finding these little creatures in real life. Although many are kept as pets, they are almost **EXTINCT** in the wild.

23

GLOSSARY

adapted (adaptation) - when a living thing has become able to survive in its surroundings by developing special features/skills over a long period of time.

burrows - tunnels or holes in the ground made by an animal.

cold-blooded - animals with bodies that do not produce their own heat, and have to be heated or cooled by their surroundings.

endangered - if a type of animal or plant is in danger of dying out forever, then they are endangered.

extinct - when a type of plant or animal no longer exists anywhere in the world.

hibernate - animals that hibernate spend all of their time sleeping during the winter and only wake up when it's warm again.

mammals - are animals with specific features. They all have hair or fur, drink milk from their mothers as babies, have a backbone, and are warm-blooded.

nocturnal - animals that sleep in the day and come out at night-time.

predators - animal that hunt and kill other animals for food.

prey - an animal that is hunted by other animals for food.

reptiles - are animals with specific features. They have dry skin with scales, a backbone, breathe using lungs and they are cold-blooded (see left).

scavengers - don't hunt live animals. They eat animals that are already dead.

venomous - poisonous, or containing poison.

The Author
Annabel Griffin is a writer and artist based in London, UK. Having worked as a bookseller for many years, she is now working in the children's publishing industry. Annabel's most recent publications include *Seasons* and *The Spectacular Lives of Sharks*.

The Illustrator
Rose Maclachlan is an illustrator based in Devon, who graduated from Falmouth University with a BA in Illustration. She likes to experiment with collage and texture to create her work and takes inspiration from her love of the outdoors and the beach.